The R.A.M.S. Library of Alchemy

Volume 3

Artephius
His Secret Book

R.A.M.S. Publishing Company

ARTEPHIUS

His Secret Book

William Salmon, Translator

Philip N. Wheeler, Editor

Produced by

Restorers of Alchemical Manuscripts Society

R.A.M.S. Publishing Company

R.A.M.S. Publishing Company
117 Rutherford Lane
Stuarts Draft VA 24477

First Edition 2015

ISBN-13 **978-1508596929**
ISBN-10 **1508596921**

Image Processing by Philip N. Wheeler

Printed in the United States of America

INTRODUCTION

Philip N. Wheeler

"His Secret Book" is attributed to Artephius, circa 1126, who was an Alchemist from Al-Andalus (the name given to Islamic Spain). Artephius was one of the most prominent Alchemists in the Middle Ages and the author of numerous works on Alchemy.

"His Secret Book" describes the entire process of preparing the Philosopher's Stone. Three separate operations are described:

1. Preparation of the Secret Fire, which is the solvent used in the Work;
2. Preparation of the mercury from antimony and iron; and
3. Preparation of the Stone itself.

The operations are not presented in order, and as in many works on Alchemy, the language is obscure: several names can refer to the same thing, and sometimes one name can refer to several things. Overall, however, it is a very clearly written text well worth further study.

Comments marked "-hwn" were made by Hans W. Nintzel.

The first part of this book was made available through R.A.M.S. as a separate publication. The second part, which starts at about page 90, is from the notebooks of Sigismund Bacstrom, M.D. It is hoped that the student can gain more insight by having both versions of the body of the text.

Dedicated to Hans W. Nintzel,

American Alchemist

and

Founder of the

Restorers of Alchemical Manuscripts Society

(R.A.M.S.)

PREFACE to "Artephius His Secret Book"[1]

The Preface to the reader in

The French and Latin copies.

Amongst all the Philosophers (loving reader) only our Artephius is not envious as he affirms himself in many places and therefore he lays down the whole Art in most open words in this Treatise. Interpretating as far as he may the doubtful speeches and sophisms of others. Nevertheless, lest he should give unto the wicked ignorant and evil men occasion and means to do hurt he has little veiled the truth in the Principles of the Science, under an artificial method, sometimes affirming, sometimes denying, and making as though he often repeated one and the same thing, whereas in those Repetitions he always changes some words, seeming often to say the contrary of what he had said before, willing to leave unto the judgement of the reader the way of truth, virtue, and true working; which if any man find let him give immortal thanks to God alone, but if he see that he works not in the right way, let him read over the author again and again, until he understand his meaning.

So did the learned John Pontanus which says in his Epistle, printed in Theatrum Chemicum. They err

---

[1] This Preface is part of a version of "His Secret Book" found in Sigismund Bacstrom's notebooks, and included in this book. -pnw

(says he speaking of them that labour in this Art) they have erred and will always err because the Philosophers in their books have never set down the proper agent except only one, which is called Artephius, but he speaks for himself and if I had not read Artephius, and understood whereof he spake, I had never come to the compliment of the work.

Therefore read this book and read it again, until you understand his speech and so obtain your desired end. It shall be needless to speak anymore concerning our Author. It suffices that by the grace of God and the use of the wonderful Quintessence he lived a 1000 years, as witnesses Roger Bacon, in his book of the wonderful works of Nature and also the most learned Theophrastus Paracelsus, in his book of Long Life; which term of 1000 years which none of the other Philosophers, no not the Father of them, Hermes himself was able to attain unto. Look therefore whether peradventure, this man have not understood the virtue of our Stone and the manner how to use it, better than the rest.

However it be, use thou it and our Labours to the glory of God and the profit of this Kingdom. Farewell.

M. B.[2] I wrote this preface.

---

[2] I don't know who M.B. is. -pnw

THE EPISTLE

OF

JOHN PONTANUS

Mentioned in the Preface of
Artephius
His Secret Book.

Wherein he bears witness of the Book translated out of
the Latin copy Extant in the third of Theatrum
Chemicum at the 775th page.

Translated out of Latin
By
William Salmon, Professor of Physick

The Epistle of John Pontanus

I, John Pontanus, have traveled through many Countrys that I might know some certainty of the Philosophers Stone, and going through as it were all the world I found many false deceivers but no true Philosophers yet continually studying and making many doubts, at the length I found the truth, but when I knew the matter in general, I yet erred two hundred times before I could attain the true matter with the operation and practice thereof.

First I began to work with the matter by putrefaction nine months together and I found nothing, then I put it into Balneum Mariae for a certain time and therein I likewise erred, afterwards I put it into the fire of calcinations for three months space and I wrought amiss, I tried all kinds of distillations and sublimations (as the Philosophers Gibor Archolaus and all the rest either say or seem to say) and I found nothing. In short I assayed to perfect the subject of the whole Art of Alchemy by all means possible to be devised by dung baths, ashes and other fires of divers kinds which are found in the Philosophers Books, but I found no good in them; wherefore I studied three whole years in the Books of the Philosophers, especially in Hermes, alone whose brief words do comprehend the whole Stone, though he spoke obscurely of the Superior and Inferior (of that which is above and that which is below) of heaven and earth therefore our instrument which brings the matter into being in the beginning, second and third work, is not the fire of a bath or dung, nor of ashes, nor of the other fires which the Philosophers have put in their Books. What fire is it then which perfects the whole work from the beginning to the ending? Surely the Philosophers have concealed it, but I being moved with pity will declare it unto you together with the compliment of the whole work.

The Philosophers Stone therefore is one, but it has many names and before you know it, it will be very difficult, for it is watery, airy, fiery, earthy, phlegmatic, choleric, and melancholy. For it is Sulphurous and it is likewise Argent-vive and it has many superfluities, which by the living God are turned into the true essence, our fire being the means and not separate anything from the subject thinking it to be necessary, he truly knows nothing at all in Philosophy, for that which is superfluous, unclean, filthy, feculent and in short the whole substance of the subject is perfected into a fixed spiritual body by the means of our fire, and this the wise men never revealed, and therefore few do come unto the Art thinking there is some superfluous or unclean thing.

Now we must see and find out the properties of our fire, and whether it agree to our matter after the manner that I have said, to wit, that it may be transmuted, whereas that fire does burn the matter, it separates nothing from the matter, it divides not the pure parts from the impure, as all the Philosophers say, but it turns the whole subject into purity. It does not sublime as Geber makes his sublimations; Arnold likewise and others speaking of sublimations and distillations to be done in a short time. It is mineral, equil, continual, it vapours not except it be too much stirred up, it partakes of Sulphur, it is taken from elsewhere yjan from the matter, it pulls down all things, it dissolves and conceals, likewise it both congeals and calcines and it is very artificial to find out and is a compendious and near way without any cost, at least with small cost, and that firing is it with a mean firing, for with a soft fire all the whole work is perfected and it performs withal, all the divers sublimations. That they should read Geber and all the other Philosophers, tho they should live an 100,000 years, could not comprehend it, because that fire is found by deep and profound meditations only, and then it may be gathered out of books and not before and therefore the error of this Art is not to find the fire which turns the whole matter into the true stone of the Philosophers, and therefore study upon it, for if I had found that first, I never erred 200 times in my practice upon the matter, wherefore I do not marvel if so many and great wise men have not attained to the work; they do err, they have erred, they will err because the

Philosophers have not put the proper agent, save only one which is named Artephius, but he speaks for himself or by himself, and unless I had read Artephius and let him speak, I had never come to the compliment of the work, but the practice is this: Let it be taken and ground with a physical confusion as diligently as may be, and let it be set upon fire, and let the proportion of the fire be known, to wit, that it only stir up the matter, and in a short time that fire without a laying on of hands will accomplish the whole work, because it will putrify, corrupt, engender and perfect and make to appear the three principle colouors, black, white and red, and by the means of our fire the medicine will be multiplied if it be joined with the crude matter, not only in quality but also in virtue; with all thy strength therefore search out this fire, and you shall attain your wish, because it does the whole work and is the key of the Philosophers, which they never revealed but if thou muse well and profoundly upon these things that have been spoken concerning the properties of the fire you may know it, otherwise not.

I being moved with pity have written these things, but that I may satisfy you fully, this fire is not transmuted with the water; these things therefore I thought it to say and to warn the prudent that they spend not their money unprofitably, but know what they ought to look after, by this means they may come to the truth of the Art and not otherwise. Farewell.

Finis.

CHAPTER I

OF THE COMPOSITION OF OUR ANTIMONIAL VINEGAR,

OR SECRET WATER.

Antimony is a mineral participating of saturnine parts, and has in all respects the nature thereof. This saturnine antimony agrees with sol, and contains in itself argent vive, in which no metal is swallowed up, except gold, and gold is truly swallowed up by this antimonial argent vive.

Without this argent vive no metal whatsoever can be whitened; it whitens laton, i.e. gold; reduceth a perfect body into its prima materia, or first matter, viz. into sulphur and argent vive, of a white color, and outshining a looking glass.

It dissolves (I say) the perfect body, which is so in its own nature; for this water is friendly and agreeable with the metals, whitening sol, because it contains in itself white or pure argent vive.

And from both these you may draw a great arcanum, viz. a water of saturnine antimony, mercurial and white; to the end that it may whiten sol, not burning, but dissolving, and afterwards congealing to the consistence or likeness of white cream.

Therefore, saith the philosopher, this water makes the body to be volatile; because after it has dissolved in it, and infrigidated, it ascends above and swims upon the surface of the water.

Take (saith he) crude leaf gold, or calcined with mercury, and put it into our vinegre, made of saturnine antimony, mercurial, and sal ammoniac, in a broad glass vessel, and four inches high or more; put it into a gentle heat, and in a short time you will see elevated a liquor, as it were oil swimming atop, much like a scum.

Gather this with a spoon or feather dipping it in; and in doing so often times a day until nothing more arises; evaporate the water with a gentle heat, i.e., the superfluous humidity of the vinegre, and there will remain the quintessence, potestates or powers of gold in the form of a white oil incombustible.

In this oil the philosophers have placed their greatest secrets; it is exceeding sweet, and of great virtue for easing the pains of wounds.

CHAPER IV[3]

OF THE OPERATIONS OF OUR ANTIMONIAL VINEGAR OR MINERAL WATER

The whole, then, of this antimonial secret is, that we know how by it to extract or draw forth argent vive, out of the body of Magnesia, not burning, and this is antimony, and a mercurial sublimate.

That is, you must extract a living and incombustible water, and then congeal, or coagulate it with the perfect body of sol, i.e. fine gold, without alloy; which is done by dissolving it into a nature [mature?] white substance of the consistency of cream, and made thoroughly white.

But first this Sol by putrefaction and resolution in this water, loseth all its light and brightness, and will grow dark and black; afterwards it will ascend above the water, and by little and little will swim upon it, in a substance of a white color.

And this is the whitening of Red Laton, to sublimate it philosophically, and to reduce it into its first matter; viz. into a white incombustible sulphur, and into a fixed argent vive.

---

[3] What happened to Chapters II and III? -pnw

And so, the fixed moisture, to wit, Gold, our Body, by the reiterating of the Liquifaction or Dissolution in this our dissolving water, is changed and reduced into a fixed Sulphur, and fixed argent vive.

Thus the perfect body of sol, resumeth life in this water; it is revived, inspired, grows, and is multiplied in its kind, as all other things are.

For in this water, it so happens, that the body compounded of two bodies, viz. sol and luna, is puffed up, swells, putrefies, is raised up, and does increase by the receiving from the vegetable and animated nature and substance.

Our water also, or vinegar aforesaid, is the Vinegar of the Mountains, i.e. of sol and luna; and therefore it is mixed with gold and silver, and sticks close to them perpetually; and the body receiveth from this water a white tincture, and shines with inestimable brightness.

Who so knows how to convert, or change the body into a medicinal white gold, may easily by the same white gold change all imperfect metals into the best or finest silver.

And this white gold is called by the philosophers "luna alba philosophorum, argentum vivum album fixum, aurum alchymiae, and fumus albus", and therefore

without this our antimonial vinegar, the aurum album
of the Philosophers cannot be made.

And because in our vinegar there is a double
substance of argentum vivum, the one from antimony,
and the other from mercury sublimated, it does give a
double weight and substance of fixed argent vive, and
also augments therein the native color, weight,
substance and tincture thereof.

CHAPTER V

OF OTHER OPERATIONS OF OUR SECRET MINERAL WATER AND ITS TINCTURE

I. Our dissolving water therefore carries with it a great tincture, and a great melting or dissolving; because that when it feels the vulgar fire, if there be in it the pure and fine bodies of sol or luna, it immediately melts them, and converts them into its white substance such as itself is, and gives to the body color, weight, and tincture.

II. In it also is a power of liquefying or melting all things that can be melted or dissolved; it is a water ponderous, viscous, precious, and worthy to be esteemed, resolving all crude bodies into their prima materia, or first matter, viz. earth and a viscous powder; that is into sulphur, and argentum vivum.

III. If therefore you put into this water, leaves, filings, or calx of any metal, and set it in a gentle heat for a time, the whole will be dissolved, and converted into a viscous water, or white oil as aforesaid.

IV. Thus it mollifies the body, and prepares for liquefaction; yea, it makes all things fusible, viz. stones and metals, and after gives them spirit and life.

V. And it dissolves all things with an admirable solution, transmuting the perfect body into a fusible medicine, melting, or liquefying, moreover fixing, and augmenting the weight and color.

VI. Work therefore with it, and you shall obtain from it what you desire, for it is the spirit and soul of sol and luna; it is the oil, the dissolving water, the fountain, the Balneum Mariae, the praeternatural fire, the moist fire, the secret, hidden and invisible fire.

VII. It is also the most acrid vinegar, concerning which an ancient philosopher saith, I besought the Lord, and he showed me a pure clear water, which I knew to be the pure vinegar, altering, penetrating, and digesting.

VIII. I say a penetrating vinegar, and the moving instrument for putrefying, resolving and reducing gold or silver into their prima materia or first matter.

IX. And it is the only agent in the universe, which in this art is able to reincrudate metallic bodies with the conservation of their species.

X. It is therefore the only apt and natural medium, by which we ought to resolve the perfect bodies of sol and luna, by a wonderful and solemn dissolution, with the conservation of the species, and without any destruction, unless it be to a new, more noble, and better form or generation, viz. into the perfect Philosopher's stone, which is their wonderful secret or arcanum.

XI. Now this water is a certain middle substance, clear as fine silver, which ought to receive the tinctures of sol and luna, so as they may be congealed, and changed into a white and living Earth.

XII. For this water needs the perfect bodies, that with them after the dissolution, it may be congealed, fixed, and coagulated into a white Earth.

XIII. But if this solution is also their coagulation, for they have one and the same operation, because one is not dissolved, but the other is congealed, nor is there any other water which can dissolve the bodies, but that which abideth with them in the matter and form.

XIV. It cannot be permanent unless it be of the nature of other bodies, that they may be made one.

XV. When therefore you see the water coagulate itself with the bodies that be dissolved therein; be assured that thy knowledge, way of working, and the work itself are true and philosophic, and that you have done rightly according to art.

CHAPTER VI

OF WHAT SUBSTANCE METALS ARE TO CONSIST

IN ORDER TO DO THIS WORK

I. Thus you see that nature has to be amended by its own like nature; that is, gold and silver are to be exalted in our water, as our water also with these bodies; which water is called the medium of the soul, without which nothing has to be done in this art.

II. It is a vegetable, mineral and animal fire, which conserves the fixed spirits of sol and luna, but destroys and conquers their bodies; for it destroys, overturns, and changes bodies and metallic forms, making them to be no bodies but a fixed spirit.

III. And it turns them into a humid substance, soft and fluid, which hath ingression and power to enter into other imperfect bodies, and to mix with them in their smallest parts, and to tinge and make them perfect.

IV. But this they could not do while they remained in their metallic forms or bodies, which were dry and hard, whereby they could have no entrance into other things, so to tinge and make perfect, what was before imperfect.

V. It is necessary therefore to convert the bodies of metals into a fluid substance; for that every tincture will tinge a thousand times more in a soft and liquid substance, than when it is in a dry one, as is plainly apparent in saffron.

VI. Therefore the transmutation of imperfect metals is impossible to be done by perfect bodies, while they are dry and hard; for which cause sake they must be brought back into their first matter, which is soft and fluid.

VII. It appears therefore that the moisture must be reverted that the hidden treasure may be revealed. And this is called the reincrudation of bodies, which is the decocting and softening them, till they lose their hard and dry substance or form; because that which is dry doth not enter into, nor tinge anything except its own body, nor tinge anything besides itself.

VIII. Therefore, the dry terrene Body doth not enter into nor tinge, except its own body, nor can it tinge except it be tinged; because (as I said before) a thick drie Earthy matter does not penetrate nor tinge, and therefore, because it cannot enter or penetrate, it can meke no alteration in the matter to be altered.

IX. For this reason it is, that gold coloreth not, until its internal or hidden spirit is drawn forth out of its bowels by this, our white water, and that it may be made altogether a spiritual substance, a white vapor, a white spirit, and a wonderful soul.

CHAPTER VII

OF THE WONDERFUL THINGS DONE BY OUR WATER

IN ALTERING AND CHANGING BODIES

I. It behooves us therefore by this our water to attenuate, alter and soften the perfect bodies, to wit sol and luna, that so they may be mixed other perfect bodies.

II. From whence, if we had no other benefit by this our antimonial water, than that it rendered bodies soft, more subtile, and fluid, according to its own nature, it would suffice.

III. But more than that, it brings back bodies to their original of sulphur and mercury, that of them we may afterwards in a little time, in less than an hour's time do that above ground which nature was a thousand years doing underground, in the mines of the earth, which is a work almost miraculous.

IV. And therefore our ultimate, or highest secret is, by this our water, to make bodies volatile, spiritual, and a tincture, or tinging water, which may have ingress or entrance into bodies.

V. For it makes bodies to be merely spirit, because it reduces hard and dry bodies, and prepares them for fusion, melting and dissolving; that is, it converts them into a permanent or fixed water. (Aqua Permanens —hwn)

VI. And so it makes of bodies a most precious and desirable oil, which is the true tincture, and the permanent fixed white water, by nature hot and moist, or rather temperate, subtile, fusible as wax, which does penetrate, sink, tinge, and make perfect the work.

VII. And this our water immediately dissolves bodies (as sol and luna) and makes them into an incombustible oil, which then may be mixed with other imperfect bodies.

VIII. It also converts other bodies into the nature of a fusible salt which the philosophers call "sal alebrot philosophorum", better and more noble than any other salt, being in its own nature fixed and not subject to vanish in fire.

IX. It is an oil indeed by nature hot, subtile, penetrating, sinking through and entering into other bodies; it is called the perfect or great elixir, and the hidden secret of the wise searchers of nature.

X. He therefore that knows this salt of sol and luna, and its generation and perfection, and afterwards how go commix it, and make it homogene with other perfect bodies, he in truth knows one of the greatest secrets of nature, and the only way that leads to perfection.

CHAPTER VIII

OF THE AFFINITY OF OUR WATER,

AND OTHER WONDERFUL THINGS DONE BY IT

I. These bodies thus dissolved by our water are called argent vive, which is not without its sulphur, nor sulphur without the fixedness of sol and luna; because sol and luna are the particular means, or medium in the form through which nature passes in the perfecting or completing thereof.

II. And this argent vive is called our esteemed and valuable salt, being animated and pregnant, and our fire, for that is nothing but fire; yet not fire, but sulphur; and not sulphur only, but also quicksilver drawn from sol and luna by our water, and reduced to a stone of great price.

III. That is to say it is a matter or substance of sol and luna, or silver and gold, altered from vileness to nobility.

IV. Now you must note that this white sulphur is the father and mother of the metals; it is our mercury, and the mineral of gold; also the soul, and the ferment; yea, the mineral virtue, and the living body; our sulphur, and our quicksilver; that is, sulphur of sulphur, quicksilver of quicksilver, and mercury of mercury.

V. The property therefore of our water is, that it melts or dissolves gold and silver, and increases their native tincture or color.

VI. For it changes their bodies from being corporeal, into a spirituality; and it is in this water which turns the bodies, or corporeal substance into a white vapor, which is a soul which is whiteness itself, subtile, hot and full of fire.

VII. This water also called the tinging or blood-color-making stone, being the virtue of the spiritual tincture, without which nothing can be done; and is the subject of all things that can be melted, and of liquefaction itself, which agrees perfectly and unites closely with sol and luna from which it can never be separated.

VIII. For it joined [joins?] in affinity to the gold and silver, but more immediately to the gold than to the silver; which you are to take special notice of.

IX. It is also called the medium of conjoining the tinctures of sol and luna with the inferior or imperfect metals; for it turns the bodies into the true tincture, to tinge the said imperfect metals, also it is the water that whiteneth, as it is whiteness itself, which quickeneth, as it is a soul; and therefore as the philosopher saith, quickly entereth into its body.

X. For it is a living water which comes to moisten the earth, that it may spring out, and in its due season bring forth much fruit; for all things springing from the earth, are endued through dew and moisture.

XI. The earth therefore springeth not forth without watering and moisture; it is the water proceeding from May dew that cleanseth the body; and like rain it penetrates them, and makes one body of two bodies.

XII. This aqua vita or water of life, being rightly ordered and disposed with the body, it whitens it, and converts or changes it into its white color.

XIII. For this water is a white vapor, and therefore the body is whitened with it.

XIV. It behoves you therefore to whiten the body, and open its unfoldings, for between these two, that is between the body and the water, there is desire and friendship, like as between male and female, because of the propinquity and likeness of their natures.

XV. Now this our second and living water is called "Azoth", the water washing the laton viz. the body compounded of sol and luna by our first water; it is also called the soul of the dissolved bodies, which souls we have even now tied together, for the use of the wise philosopher.

XVI. How precious then, and how great a thing is this water; for without it, the work could never be done or perfected; it is also called the "vase naturae", the belly, the womb, the receptacle of the tincture, the earth, the nurse. (see Hermes —hwn)

XVII. It is the royal fountain in which the king and queen bathe themselves; and the mother must be put into and sealed up within the belly of her infant; and that is sol himself, who proceeded from her, and whom she brought forth; and therefore they have loved one another as mother and son, and are conjoined together, because they come from one and the same root, and are of the same substance and nature.

XVIII. And because this water is the water of the vegetable life, it causes the dead body to vegetate, increase and spring forth, and to rise from death to life, by being dissolved first and then sublimed.

XIX. And in doing this the body is converted into a spirit, and the spirit afterwards into a body; and then is made the amity, the peace, the concord, and the union of contraries, to wit, between the body and the spirit, which reciprocally, or mutually change their natures which they receive, and communicate one to another through their most minute parts.

XX. So that that which is hot is mixed with that which is cold, the dry with the moist, and the hard with the soft; by which means, there is a mixture made of contrary natures, viz. of cold and hot, and moist with dry, even most admirable unity between enemies.

CHAPTER IX

OF SUBLIMATION: OR, THE SEPARATION OF THE PURE,

FROM THE IMPURE, BY THIS WATER

I. Our dissolution then of bodies, which is made such in this first water, is nothing else, but a destroying or overcoming of the moist with the dry, for the moist is coagulated with the dry.

II. For the moisture is contained under, terminated with, and coagulated in the dry body, to wit, in that which is earthy.

III. Let therefore the hard and the dry bodies be put into our first water in a vessel, which close well, and let them there abide till they be dissolved, and ascend to the top; then may they be called a new body, the white gold made by art, the white stone, the white sulphur, not inflammable, the paradisical stone, viz. the stone transmuting imperfect metals into fine white silver.

IV. Then we have also the body, soul and spirit altogether; of which spirit and soul it is said, that they cannot be extracted from the perfect bodies, but by the help or conjunction of our dissolving water.

V. Because it is certain, that the things fixed cannot be lifted up, or made to ascend, but by the conjunction or help of that which is volatile.

VI. The spirit, therefore, by help of the water and the soul, is drawn forth from the bodies themselves, and the body is thereby made spiritual; for that at the same instant of time, the spirit, with the soul of the bodies, ascends on high to the superior part, which is the perfection of the stone and is called sublimation.

VII. This sublimation, is made by things acid, spiritual, volatile, and which are in their own nature sulphureous and viscous, which dissolves bodies and makes them to ascend, and be changed into air and spirit.

VIII. And in this sublimation, a certain part of our said first water ascends with the bodies, joining itself with them, ascending and subliming into one neutral and complex substance, which contains the nature of the two, viz. the nature of the two bodies and of the water.

IX. And therefore it is called the corporeal and spiritual compositum, corjufle, cambar, ethelia, zandarith, duenech, the good; but properly it is called the permanent or fixed water only, because it flies not in the fire.

X. But it perpetually adheres to the commixed or compound bodies, that is, the sol and luna, and communicates to them the living tincture, incombustible and most fixed, much more noble and precious than the former which these bodies had.

XI. Because from henceforth this tincture runs like oil, running through and penetrating bodies, and giving to them its wonderful fixity; and this tincture is the spirit, and the spirit is the soul, and the soul is the body.

XII. For in this operation, the body is made a spirit of a most subtile nature; and again, the spirit is corporified and changed into the nature of the body, with the bodies, whereby our stone consists of a body, a soul, and a spirit.

XIII. O God, how through nature, doth thou change a body into a spirit: which could not be done, if the spirit were not incorporated with the bodies, and the bodies made volatile with the spirit, and afterwards permanent and fixed.

XIV. For this cause sake, they have passed over into one another, and by the influence of wisdom, are converted into one another. O Wisdom: how thou makest the most fixed gold to be volatile and fugitive, yeah, though by nature it is the most fixed of all things in the world!

XV. It is necessary therefore, to dissolve and liquefy these bodies by our water, and to make them a permanent or fixed water, a pure, golden water leaving in the bottom the gross, earthy, superfluous and dry matter.

XVI. And in this subliming, making thin and pure, the fire ought to be gentle; but if in this subliming with soft fire, the bodies be not purified, and the gross

and earthy parts thereof (note this well) be not
separated from the impurities of the Dead, you shall
not be able to perfect the work. For thou needest
nothing but the thin and subtile part of the dissolved
bodies, which our water will give thee, if thou
proceedest with a slow or gentle fire, by separating
the things heterogene from the things homogene.

CHAPTER X

OF THE SEPARATION OF THE PURE PARTS FROM THE IMPURE

I. This compositum then has its mundification or cleaning, by our moist fire, which by dissolving and subliming that which is pure and white, it cast forth its feces or filth like a voluntary vomit.

II. For in such a dissolution and natural sublimation or lifting up, there is a loosening or untying of the elements, and a cleansing and separating of the pure from the impure.

III. So that the pure and white substance ascends upwards and the impure and earthy remains fixed in the bottom of the water and the vessel.

IV. This must be taken away and removed, because it is of no value, taking only the middle white substance, flowing and melted or dissolved, rejecting the feculent earth, which remains below in the bottom.

V. These feces were separated partly by the water, and are the dross and terra damnata, which is of no value, nor can do any such service as the clear, white, pure and clear matter, which is wholly and only to be taken and made use of.

VI. And against this capharean rock, the ship of knowledge, or art of the young philosopher is often, as it happened also to me sometimes, dashed together in pieces, or destroyed, because the philosophers for the most part speak by the contraries.

VII. That is to say that nothing must be removed or taken away, except the moisture, which is the blackness; which notwithstanding they speak and write only to the unwary, who, without a master, indefatigable reading, or humble supplications to God Almighty, would ravish away the golden fleece.

VIII. It is therefore to be observed, that this separation, division, and sublimation, is without a doubt the key to the whole work.

IX. After the putrefaction, then, and dissolution of these bodies, our bodies also ascend to the top, even to the surface of the dissolving water, in a whiteness of color, which whiteness is life.

X. And in this whiteness, the antimonial and mercurial soul, is by natural compact infused into, and joined with the spirits of sol and luna, which separate the thin from the thick, and the pure from the impure.

XI. That is, by lifting up, by little and little, the thin and the pure part of the body, from the feces and impurity, until all the pure parts are separated and ascended.

XII. And in this work is out natural and philosophical sublimation work completed.

XIII. Now in this whiteness is the soul infused into the body, to wit, the mineral virtue, which is more subtile than fire, being indeed the true quintessence and life, which desires or hungers to be born again, and to put off the defilements and be spoiled of its gross and earthy feces, which it has taken from its monstrous womb, and corrupt place of its original.

XIV. And in this our philosophical sublimation, not in the impure, corrupt, vulgar mercury, which has no qualities or properties like to those, with which our mercury, drawn from its vitriolic caverns is adorned. But let us return to our sublimation.

CHAPTER XI

OF THE SOUL WHICH IS EXTRACTED BY OUR WATER,

AND MADE TO ASCEND

I. It is most certain therefore in this art, that this soul extracted from the bodies, cannot be made to ascend, but by adding to it a volatile matter, which is of its own kind.

II. By which the bodies will be made volatile and spiritual, lifting themselves up, subtilizing and subliming themselves, contrary to their own proper nature, which is corporeal, heavy and ponderous.

III. And by this means they are unbodied, or made no bodies, to wit, incorporeal, and a quintessence of the nature of a spirit, which is called, "avis hermetis", and "mercurius extractus", drawn from a red subject or matter.

IV. And so the terrene or earthy parts remain below, or rather the grosser parts of the bodies, which can by no industry or ingenuity of man be brought to a perfect dissolution.

V. And this white vapor, this white gold, to wit, this quintessence, is called also the compound magnesia, which like a man does contain, or like a man is composed of a body, soul and spirit.

VI. Now the body is the fixed solar earth, exceeding the most subtile matter, which by the help of our divine water is with difficulty lifted up or separated.

VII. The soul is the tincture of sol and luna, proceeding from the conjunction, or communication of these two, to wit, the bodies of sol and luna, and our water.

VII. And the spirit is the mineral power, or virtue of the bodies, and also out of the bodies like as the tinctures or colors in dying cloth are by the water put upon, and diffused in and through the cloth.

VIII. And this mercurial spirit is the chain or band of the solar soul; and the solar body is that body which contains the spirit and soul, having the power of fixing in itself, being joined with luna.

IX. The spirit therefore penetrates, the body fixes, and the soul joins together, tinges and whitens. From these three bodies united together is our stone made: to wit, sol, luna and mercury.

X. The Spirit therefore penetrates, the Body fixes and the Soul joynes together, tinges and whitens.

XI. From these three united together, is our Stone made; to wit, of Sol, Luna and Mercury.

XII. Therefore with this our golden water, a natural substance is extracted, exceeding all natural substances; and so, except the bodies be broken and destroyed, imbibed, made subtile and fine, thriftily, and diligently managed, till they are abstracted from, or lose their grossness or solid substance, and be changed into a subtile spirit, all our labor will be in vain.

XIII. And unless the bodies be made no bodies or incorporeal, that is converted into the philosophers mercury, there is no rule of art yet found out to work by.

XIV. The reason is, because it is impossible to draw out of the bodies all that most thin and subtile spirit, which has in itself the tincture, except it first be resolved in our water.

XV. Dissolve then the bodies in this our golden water, and boil them until all the tincture is brought forth by the water, in a white color and a white oil; and when you see this whiteness upon the water, then know that the bodies are melted, liquified or dissolved.

XVI. Continue then this boiling, till the dark, black, and white cloud is brought forth, which they have conceived.

CHAPTER XII

OF DIGESTION, AND HOW THE SPIRIT IS MADE THEREBY

I. Put therefore the perfect bodies of metals, to wit, sol and luna, into our water in a vessel, hermetically sealed, upon a gentle fire, and digest continually, till they are perfectly resolved into a most precious oil.

II. Digest (saith Adfar) with a gentle fire, as it were for the hatching of chickens, so long till the bodies are dissolved, and their perfectly conjoined tincture (mark this well) is extracted.

III. But it is not extracted all at once, but it is drawn out by little and little, day by day, and hour by hour, till after a long time, the solution thereof is completed, and that which is dissolved always swims on top.

IV. And while this dissolution is in hand, let the fire be gentle and continual, till the bodies are dissolved into a viscous and most subtile water, and the whole tincture be educed, in color first black, which is the sign of a true dissolution.

V. Then continue the digestion, till it become a white fixed water, for being digested in balneo, it will afterwards become clear, and in the end become like common argent vive, ascending by the spirit above the first water.

VI. When there you see bodies dissolved in the first viscous water, then know, that they are turned into a vapor, and that the soul is separated from the dead body, and by sublimation, turned into the order of spirits.

VII. Whence both of them, with a part of our water, are made spirits flying up in the air; and there the compounded body, made of the male and female, viz. of sol and luna, and of that most subtile nature, cleansed by sublimation, taketh life, and is made spiritual by its own humidity.

VIII. That is by its own water; like as a man is sustained by the air, whereby from thenceforth it is multiplied, and increases in its own kind, as do all other things.

IX. In such an ascension therefore, and philosophical sublimation, all are joined one with another, and the new body subtilized, or made living by the spirit, miraculously liveth or springs like a vegetable.

X. Wherefore, unless the bodies be attenuated, or made thin, by the fire and water, till they ascend in a spirit, and are made or do become like water and vapor or mercury, you labor wholly in vain.

XI. But when they arise or ascend, they are born or brought forth in the air or spirit, and in the same they are changed, and made life with life, so as they can never be separated, but are as water mixed with water.

XII. And therefore, it is wisely said, that the stone is born of the spirit, because it is altogether spiritual.

XIII. For the vulture himself flying without wings cries upon the top of the mountain, saying, I am the white brought forth from the black, and the red brought forth from the white, the citrine son of the red; I speak the truth and lie not.

CHAPTER XIII

OF THE BEGINNING OF THE WORK,

AND A SUMMARY OF WHAT IS TO BE DONE

I. It sufficeth thee then to put the bodies in the vessel, and into the water once and for all, and to close the vessel well, until a true separation is made.

II. This the obscure artist calls conjunction, sublimation, assation, extraction, putrefaction, ligation, desponsation, subtilization, generation, etc.

III. Now the whole magistery may be perfected, work, as in the generation of man, and of every vegetable; put the seed once into the womb, and shut it up well.

IV. Thus you may see that you need not many things, and that this our work requires no great charges, for that there is but one stone, there is but one medicine, one vessel, one order of working, and one successive disposition to the white and to the red.

V. And although we say in many places, take this, and take that, yet we understand, that it behoves us to take but one thing, and put it once into the vessel, until the work be perfected.

VI. But these things are so set down by obscure philosophers to deceive the unwary, as we have before spoken; for is not this "ars cabalistica" or a secret and a hidden art? Is it not an art full of secrets? And believest thou O fool that we plainly teach this secret of secrets, taking our words according to their literal signification?

VII. Truly, I tell thee, that as for myself, I am no ways self seeking, or envious as others are; but he that takes the words of the other philosophers according to their common signification, he even already, having lost Ariadne's clue of thread, wanders in the midst of the labyrinth, multiplies errors, and casts away his money for naught.

VIII. And I, Artephius, after I became an adept, and had attained to the true and complete wisdom, by studying the books of the most faithful Hermes, the speaker of truth, was sometimes obscure also as others were.

IX. But when I had for the space of a thousand years, or thereabouts, which has now passed over my head, since the time I was born to this day, through the alone goodness of God Almighty, by the use of this wonderful quintessence.

X. When I say for so very long a time, I found no man had found out or obtained this hermetic secret, because of the obscurity of the philosophers words.

XI. Being moved with a generous mind, and the integrity of a good man, I have determined in these latter days of my life, to declare all things truly and sincerely, that you may not want anything for the perfecting of this stone of the philosophers.

XII. Excepting one certain thing, which is not lawful for me to discover to any, because it is either revealed or made known by God himself, or taught by some master, which notwithstanding he that can bend himself to the search thereof, by the help of a little experience, may easily learn in this book.

XIII. In this book I have therefore written the naked truth, though clothed or disguised with few colors; yet so that every good and wise man may happily gather these desirable apples of the Hesperides from this our philosophers tree.

XIV. Wherefore praises be given to the most high God, who has poured into our soul of his goodness; and through a good old age, even an almost infinite number of years, has truly filled our hearts with his love, in which, methinks, I embrace, cherish, and truly love all mankind together.

XV. But to return to out business. Truly our work is perfectly performed; for that which the heat of sun is a hundred years in doing, for the generation of one metal in the bowels of the earth; our secret fire, that is, our fiery and sulphureous water, which is called Balneum Mariae (!! —hwn), doth as I have often seen in a very short time.

CHAPTER XIV

OF THE EASINESS AND SIMPLICITY OF THIS WORK,

AND, OF OUR PHILOSOPHIC FIRE

I. Now this operation or work is a thing of no great labor to him who knows and understands it; nor is the matter so dear (considering how small a quantity does suffice) that it may cause any man to withdraw his hand from it.

II. It is indeed, a work so short and easy, that it may well be called woman's work, and the play of children.

III. Go to it then,, my son, put up thy supplications to God almighty; be diligent in searching the books of the learned in this science; for one book openeth another; think and meditate of these things profoundly; and avoid all things which vanish in or will not endure the fire, because from these adjustible, perishing or consuming things, you can never attain to the perfect matter, which is only found in the digesting of your water, extracted from sol and luna.

IV. For by this water, color, and ponderosity or weight, are infinitely given to the matter; and this water is a white vapor, which like a soul flows through the perfect bodies, taking wholly from them their blackness, and impurities, uniting the two bodies in one, and increasing their water.

V. Nor is there any other thing than Azoth, to wit, this our water, which can take from the perfect bodies of sol and luna, their natural color, making the red body white, according to the disposition thereof.

VI. Now let us speak of the fire. Our fire is mineral, equal, continuous; it fumes not, unless it be too much stirred up, participates of sulphur, and is taken from other things than from the matter; it overturns all things, dissolves, congeals, and calcines, and is to be found out by art, or after an artificial manner.

VII. It is a compendious thing, got without cost or charge, or at least without any great purchase; it is humid, vaporous, digestive, altering, penetrating, subtile, spiritous, not violent, incombustible, circumspective, continent, and one only thing.

VIII. It is also a fountain of living water, which circumvolveth and contains the place, in which the king and queen bathe themselves; through the whole work this moist fire is sufficient; in the beginning, middle and end, because in it, the whole of the art does consist.

IX. This is the natural fire, which is yet against nature, not natural and which burns not; lastly, this fire is hot, cold, dry, moist; meditate on these things and proceed directly without anything of a foreign nature.

X. If you understand not these fires, give ear to what I have yet to say, never as yet written in any book, but drawn from the more abstruse and occult riddles of the ancients.

CHAPTER XV

OF THE THREE KINDS OF FIRES OF THE

PHILOSOPHERS IN PARTICULAR

I. We have properly three fires, without which our art cannot be perfected; and whosoever works without them takes a great deal of labor in vain.

II. The first fire is that of the lamp, which is continuous, humid, vaporous, spiritous, and found out by art.

III. This lamp ought to be proportioned to the enclosure; wherein you must use great judgement, which none can attain to, but he that can bend to the search thereof.

IV. For if this fire of the lamp be not measured, or duly proportioned or fitted to the furnace, it will be, that either for the want of heat you will not see the expected signs, in their limited times, whereby you will lose your hopes and expectation by a too long delay; or else, by reason of too much heat, you will burn the "flores auri", the golden flowers, and so foolishly bewail your lost expense.

V. The second fire is ignis cinerum, an ash heat, in which the vessel hermetically sealed is recluded, or buried; or rather it is that most sweet and gentle heat, which proceeding from the temperate vapors of the lamp, does equally surround your vessel.

VI. This fire is not violent or forcing, except it be too much excited or stirred up; it is a fire digestive; alterative, and taken from another body than the matter; being but one only, moist also, and not natural.

VII. The third fire, is the natural fire of water, which is also called the fire against nature, because it is water; and yet nevertheless, it makes a mere spirit of gold, which common fire is not able to do.

VIII. This fire is mineral, equal, and participates of sulphur; it overturns or destroys, congeals, dissolves, and calcines; it is penetrating, subtile, incombustible and not burning, and is the fountain of living water, wherein the king and queen bathe

<u>themselves</u>, whose help we stand in need of through the whole work, through the beginning, middle, and end.

IX. But the other two above mentioned, we have not always occasion for, but only at some times.

X. In reading therefore the books of the philosophers, conjoin these three fires in your judgement, and without doubt, you will understand whatever they have written of them.

CHAPTER XVI

OF THE COLOURS OF OUR PHILOSOPHICK TIMCTURE, OR STONE

I. Now as to the colors, that which does not make black cannot make white, because blackness is the beginning of whiteness, and a sign of putrefaction and alteration, and that the body is now penetrated and mortified.

II. From the putrefaction therefore in this water, there first appears blackness, like unto broth wherein some bloody thing is boiled.

III. Secondly, the black earth by continual digestion is whitened, because the soul of the two bodies swims above upon the water, like white cream; and in this only whiteness, all the spirits are so united, that they can never fly one from another.

IV. And therefore the latten must be whitened, and its leaves unfolded, i.e., its body broken or opened, lest we labor in vain; for this whiteness is the perfect stone for the white work, and a body ennobled to that end; even a tincture of a most exuberant glory, and shining brightness, which never departs from the body it is once joined with.

V. Therefore you must note here, that the spirits are not fixed but in the white color, which is more noble than the other colors, and is more vehemently to be desired, for that as it were the complement or perfection of the whole work.

VI. For our earth putrefies and becomes black, then it is putrefied in lifting up or separation; afterwards being dried, its blackness goes away from it, and then it is whitened, and the feminine dominion of the darkness and humidity perisheth; then also the white vapor penetrates through the new body, and the spirits are bound up or fixed in the dryness.

VII. And that which is corrupting, deformed and black through the moisture, vanishes away; so the new body rises again clear, pure, white and immortal, obtaining the victory over all its enemies.

VIII. And as heat working upon that which is moist, causeth or generates blackness, which is the prime or first color, so always by decoction more and more heat working upon that which is dry begets whiteness, which is the second color; and then working upon that which is purely and perfectly dry, it produces citrinity and redness, thus much for colors.

IX. We must know therefore, that thing which has its head red and white, but its feet white and afterwards red; and its eyes beforehand black, that this thing, I say, is the only matter of our magistery.

CHAPTER XVII

OF THE PERFECT BODIES, THEIR PUTREFACTION, CORRUPTION, DIGESTION AND TINCTURE

I. Dissolve then sol and luna in our dissolving water, which is familiar and friendly, and next in nature to them; and is also sweet and pleasant to them, and as it were a womb, a mother, an original, the beginning and the end of their life.

II. That is the reason why they are meliorated or amended in this water, because like nature, rejoices in like nature, and like nature retains like nature, being joined the one to the other, in a true marriage, by which they are made one nature, one new body, raised again from the dead, and immortal.

III. Thus it behoves you to join consanguinity, or sameness of kind, by which these natures, will meet and follow one another, purify themselves and generate, and make one another rejoice; for that like nature now is disposed by like nature, even that which is nearest, and most friendly to it.

IV. Our water then is the most beautiful, lovely, and clear fountain, prepared only for the king, and queen whom it knows very well, and they it.

V. For it attracts them to itself, and they abide therein for two or three days, to wit, two or three months, to wash themselves therewith, whereby they are made young again and beautiful.

VI. And because sol and luna have their original from this water their mother; it is necessary therefore that they enter into it again, to wit, into their mother's womb, that they may be regenerated and born again, and made more healthy, more noble and more strong.

VII. If therefore these do not die and be converted to water, they remain alone or as they were and without fruit; but if they die, and are resolved in our water, they bring forth fruit of a hundred fold; and from that very place in which they seem to perish, from thence shall they appear to be that which they were not before.

VIII. Let therefore the spirit of our living water be, with all care and industry, fixed with sol and luna; for they being converted into the nature of water become dead, and appear like to the dead; from thence afterwards being revived, they increase and multiply, even as do all sorts of vegetable substances.

IX. It suffices then to dispose the matter sufficiently without, because that within, it sufficiently disposes itself for the perfection of its work.

X. For it has in itself a certain and inherent motion, according to the true way and method, and a much better order than it is possible for any man to invent or think of.

XI. For this cause it is that you need only prepare the matter, nature herself will perfect it; and if she be not hindered by some contrary thing, she will not overpass her own certain motion, neither in conceiving or generating, nor in bringing forth.

XII. Wherefore, after the preparation of the matter, beware only lest by too much heat or fire, you inflame the bath, or make it too hot; secondly, take heed lest the spirit should exhale, lest it hurt the operator, to wit, lest it destroy the work, and induce many informities, as trouble, sadness, vexation, and discontent.

XIII. From these things which have been spoken, this axiom is manifest, to wit, that he can never know the necessary course of nature, in the making or generating of metals, who is ignorant of the way of destroying them.

XIV. You must therefore join them together that are of one consanguinity or kindred; for like natures do find out and join with their like natures, and by putrifying themselves, and mix together and mortify themselves.

XV. It is needful therefore to know this corruption and generation, and the natures themselves do embrace one another, and are brought to a fixity in a slow and gentle fire; how like natures rejoiceth with like natures; and how they retain one another and are converted into a white consistency.

XVI. This white substance, if you will make it red, you must continually decoct it in a dry fire till it be rubified, or become red as blood, which is nothing but water, fire, and true tincture.

XVII. And so by a continual dry fire, the whiteness is changed, removed, perfected, made citrine, and still digested till it become to a true red and fixed color.

XVIII. And consequently by how much more it is heightened in color, and made a true tincture of perfect redness.

XIX. Wherefore with a dry fire, and a dry calcination, without any moisture, you must decoct this compositum, till it be invested with a most perfect red color, and then it will be the true and perfect elixir.

CHAPTER XVIII

OF THE MULTIPLICATION OF THE PHILOSOPHICK TINCTURE

I. Now if afterwards you would multiply your tincture, you must again resolve that red, in new and fresh dissolving water, and then by decoctions first whiten, and then rubify it again, by the degrees of fire, reiterating the first method of operating in this work.

II. Dissolve, coagulate, and reiterate the closing up, the opening and multiplying in quantity and quality at your own pleasure.

III. For by a new corruption and generation, there is introduced a new motion.

IV. Thus we can never find an end if we do always work by reiterating the same thing over and over again, viz. by solution and coagulation, by the help of our dissolving water, by which we dissolve and congeal, as we have formerly said, in the beginning of the work.

V. Thus also is the virtue thereof increased, and multiplied both in quantity and quality; so that if after the first course of the operation you obtain a hundred-fold; by the second course you will have a thousand-fold; and by the third, ten-thousand fold.

VI. And by pursuing your work, your projection will come to infinity, tinging truly and perfectly, and fixing the greatest quantity how much soever.

VII. Thus by a thing of small and easy price, you have both color, goodness, and weight.

VIII. Our fire then and azoth are sufficient for you: decoct, reiterate, dissolve, congeal, and continue this course, according as you please, multiplying it as you think good, until your medicine is made fusible as wax, and has attained the quantity and goodness or fixity and color you desire.

IX. This then is the compleating of the whole work of our second stone (observe it well) that you take the perfect body, and put it into our water in a glass vesica or body well closed, lest the air get in or the enclosed humidity get out.

X. Keep it in digestion in a gentle heat, as it were of a balneum, and assiduously continue the operation or work upon the fire, till the decoction and digestion is perfect.

XI. And keep it in this digestion of a gentle heat, until it be purified and re-solved into blackness, and be drawn up and sublimed by the water, and is thereby cleaned from all blackness and impurity, that it may be white and subtile.

XII. Until it comes to the ultimate or highest purity of sublimation, and utmost volatility, and be made white both within and without: for the vulture flying in the air without wings, cries out that it might get up upon the mountain, that is upon the waters, upon which the "spiritus albus" or spirit of whiteness is born.

XIII. Continue still a fitting fire, and that spirit, which is the subtile being of the body, and of the mercury will ascend upon the top of the water, which quintessence is more white than the driven snow.

XIV. Continue yet still, and towards the end, increase the fire, till the whole spiritual substance ascend to the top.

XV. And know well, that whatsoever is clear, white-pure and spiritual, ascends in the air to the top of the water in the substance of a white vapor, which the philosophers call their Virgins Milk.

CHAPTER XIX

OF SUBLIMATION IN PARTICULAR, AND

SEPARATION OF THE PURE FROM THE IMPURE

I. It ought to be, therefore, as one of the Sybills said, that the son of the virgin be exalted from the earth, and that the white quintessence after its rising out of the dead earth, be raised up towards heaven; the gross and thick remaining in the bottom, of the vessel and the water.

II. Afterwards, the vessel being cooled, you will find in the bottom the black feces, scorched and burnt, which separate from the spirit and quintessence of whiteness, and cast them away.

III. Then will the argent vive fall down from our air and spirit, upon the new earth, which is called argent vive sublimed by the air or spirit, whereof is made a viscous water, pure and white.

IV. This water is the true tincture separated from all its black feces, and our brass or latten is prepared with our water, purified and brought to a white color.

V. Which white color is not obtained but by decoction and coagulation of the water; decoct, therefore, continually, wash away the blackness from the latten, not with your hands, but with the stone, or the fire, or our second mercurial water which is the true tincture.

VI. This separation of the pure from the impure is not done with hands, but nature herself does it, and brings it to perfection by a circular operation.

VII. It appears then, that this composition is not a work of hands, but a change of the natures; because nature dissolves and joins itself, sublimes and lifts itself up, and grows white, being separated from the feces.

VIII. And in such a sublimation the more subtile, pure, and essential parts are conjoined; for that with the fiery nature or property lifts up the subtile parts, it separates always the more pure, leaving the grosser at the bottom.

IX. Wherefore your fire ought to be gentle and a continual vapor, with which you sublime, that the matter may be filled with spirit from the air, and live.

X. For naturally all things take life from the inbreathing of the air; and so also our magistery receives in the vapor or spirit, by the sublimation of the water.

XI. Our bras or latten then, is to be made to ascend by the degrees of fire, but of its own accord, freely, and without violence; except the body therefore be by the fire and water broken, or dissolved, and attenuated, until it ascends as a spirit, or climbs like argent vive, or rather as the white soul, separated from the body, and by sublimation diluted or brought into a spirit, nothing is or can be done.

XII. But when it ascends on high, it is born in the air or spirit, and is changed into spirit; and becomes life with life, being only spiritual and incorruptible.

XIII. And by such an operation it is that the body is made spirit, of a subtile nature, and the spirit is incorporated with the body, and made one with it; and by such a sublimation, conjunction, and raising up, the whole, both body and spirit are made white.

CHAPTER XX

OF DIGESTION, SUBLIMATION, AND SEPARATION

OF THE BODIES, FOR THE PERFECTION OF THE WORK

I. This philosophical and natural sublimation therefore is necessary which makes peace between, or fixes the body and spirit, which is impossible to be done otherwise, than in the separation of these parts.

II. Therefore it behooves you to sublime both, that the pure may ascend, and the impure may descend, or be left at the bottom, in the perplexity of a troubled sea.

III. And for this reason it must be continually decocted, that it may be brought to a subtile property, and the body may assume, and draw to itself the white mercurial soul, which it naturally holds, and suffers not to be separated from it, because it is like to it in the nearness of the first pure and simple nature.

IV. From these things it is necessary, to make a separation by decoction, till no more remains of the purity of the soul, which is not ascended and exalted to the higher part, whereby they will both be reduced to an equality of properties, and a simple pure whiteness.

V. The vulture flying through the air, and the toad creeping upon the ground, are the emblems of our magistery.

VI. When therefore gently and with much care, you separate the earth from the water, that is from the fire, and the thin from the thick, then that which is pure will separate itself from the earth, and ascend to the upper part, as it were into heaven, and the impure will descend beneath, as to the earth.

VII. And the more subtile part in the superior place will take upon it the nature of a spirit, and that in the lower place, the nature of an earthy body.

VIII. Wherefore, let the white property with the more subtile part of the body, be by this operation, made to ascend leaving the feces behind, which is done in a short time.

IX. For the soul is aided by her associate and fellow, and perfected by it.

X. My mother, saith the body, has begotten me, and by me she herself is begotten; now after I have taken from her, her flying she after an admirable manner becomes kind and nourishing, and cherishing the son whom she has begotten till he come to a ripe or perfect age.

CHAPTER XXI

OF THE SECRET OPERATION OF THE WATER AND

SPIRIT UPON THE BODY

I. Hear now this secret: keep the body in our mercurial water, till it ascends with the white soul, and the earthy part descends to the bottom, which is called the residing earth.

II. Then you shall see the water coagulate itself with the body, and be assured the art is true; because the body coagulates the moisture into dryness, like as the rennet of a lamb or calf turns milk into cheese.

III. In the same manner the spirit penetrates the body, and is perfectly comixed with it in its smallest atoms, and the body draws to itself his moisture, to wit, its white soul, like as the loadstone draws iron, because of the nearness and likeness of its nature; and then one contains the other.

IV. And this is the sublimation and coagulation, which retaineth every volatile thing, making it fixed forever.

V. This compositum then is not a mechanical thing, or a work of the hands, but as I said, a changing of natures; and a wonderful connection of their cold with hot, and the moist with the dry; the hot is mixed with the cold, and the dry with the moist.

VI. By this means is made the mixture and conjunction of body and spirit, which is called a conversion of contrary spirits and natures, because by such a dissolution and sublimation, the spirit is converted into a body and body in a spirit.

VII. So that the natures being mixed together, and reduced into one, do change one another: and as the body corporifies the spirit, or changes it into a body, so also does the spirit convert the body into a tinging and white spirit.

VIII. Wherefore as the last time I say, decoct the body in our white water, viz. mercury, till it is dissolved into blackness, and then by continual decoction, let it be deprived of the same blackness, and the body so dissolved, will at length ascend or rise with a white soul.

IX. And then the one will be mixed with the other, and so embrace one another that it shall not be possible any more to separate them, but the spirit, with a real agreement, will be unified with the body, and make one permanent or fixed substance.

X. And this is the solution of the body, and coagulation of the spirit, which have one and the same operation.

XI. Whoso therefore, knows how to conjoin the principles, or direct the work, to impregnate, to mortify, to putrefy, to generate, to quicken the species, to make white, to cleanse the culture from its blackness and darkness, till he is purged by the fire and tinged, and purified from all his spots, shall be the possessor of a treasure so great that even kings themselves shall venerate him.

CHAPTER XXII

OF THE SIGNS OF THE END OF THE WORK,

AND THE PERFECTION THEREOF

I. Wherefore, let our body remain in the water till it is dissolved into a subtile powder in the bottom of the vessel and the water, which is called the black ashes; this is the corruption of the body which is called by the philosophers or wise men, "Saturnus plumbum philosophorum", and pulvis discontinuatus, viz. saturn, latten or brass, the lead of the philosophers the disguised powder.

II. And in this putrefaction and resolution of the body, three signs appear, viz., a black color, a discontinuity of parts, and a stinking smell, not much unlike to the smell of a vault where dead bodies are buried.

III. These ashes then are those of which the philosophers have spoken so much which remained in the lower part of the vessel, which we ought not to undervalue or despise.

IV. In them is the **royal diadem**, and the black and unclean argent vive, which ought to be cleansed from its blackness, by a continual digestion in our water, till it be elevated above in a white color, which is called the gander, and the bird of Hermes.

V. He therefore that maketh the red earth black, and then renders it white, has obtained the magistery. So also he who kills the living, and revives the dead.

VI. Therefore make the black white, and the white black, and you perfect the work.

VII. And when you see the true whiteness appear, which shineth like a bright sword, or polished silver, know that in that whiteness there is redness hidden.

VII. But then beware that you take not that whiteness out of the vessel, but only digest it to the end, that with heat and dryness, it may assume a citron color, and a most beautiful redness.

VIII. Which when you see, render praises and thanksgiving to the most great and good God, who gives wisdom and riches to whomsoever He pleases, and takes them away according to the wickedness of a person. To Him, I say, the most wise and almighty God, be glory for ages and ages.

AMEN.

-Finis-

ARTEPHIUS HIS SECRET BOOK

The text from the Notebooks of

Sigismund Bacstrom, M.D.

Antimony is of the parts of Saturn and has in every respect the nature thereof, so this Saturnian Antimony agrees with the Sun, having in itself Agrent-vive, wherein no metal is drowned but Gold, that is to say, Gold only is drowned in Antimonial Saturnino Argent-vive and without that Argent-vive, no metal can be whitened, it whitens therefore Laton, that is Gold; and reduces a perfect body into its First Matter; that is into Sulphur and Argent-vive, of a white colour, and shining more than glass. It dissolves, I say, the perfect body which is of his nature, for this water is friendly and pleasant to the metals, whitening the Sun because it contains a white Argent-vive, and from hence thou may draw a great secret. To wit that the water of Saturnino Antimony ought to be Mercurial and white to the end that it may whiten Gold, not burning it, but dissolving; and afterwards congealing it to the form of white Cream, therefore says the Philosopher that this water makes the body to be volatile, because after it has been dissolved in this water and cooled again, it mounts aloft on the surface of the water; take (says he) Gold crude foliated, laminated or calcined with Mercury and put it into our Vinegar, Antimonial Saturnino Mercurial and drawn from Sal ammoniac (as it is said) in a broad vessel of glass, four fingers high or more, and leave it there; in a temperate heat, and in a short time, you will see lifted up, as it were a liquor of Oil, swimming aloft in a manner of a thin skin, that gather with a spoon or with a feather dipping in it and so doing many times in a day, until there do nothing more arise, afterwards make the water vapour away, by the fire: that is to say, the superfluous humour of the Vinegar, and there will remain unto you a Fifth Essence of Gold, in form of a white Oil incombustible, wherein the Philosophers have placed their greatest secrets, and this Oil is exceeding sweet and is of great power to mitigate the pain and grief of wounds.

All the secret then of this secret Antimonials, is that by virtue thereof we know how to extract and draw out of the body of the Magnessio. Argent-vive not burning (and this Antimony and Mercurial Sublimate) that is we must draw a water living incombustible, and then congeal it to the or with the perfect body of the Sun, which is dissolved therein into a nature and substance white, congealed, as if it were Cream, which makes it all to become white.

Nevertheless, first of all this Sun in his putrefaction and resolution in this water, in the beginning will loose his light, be darkened, and become black and afterward will lift himself upon upon the water and there will swim upon it: By little and little a white colour in a white substance and this is called to whiten the Red Laton.

To sublime it Philosophically and to reduce it into his First Matter, that is to say, into white Sulphur incombustible and into Argent-vive fixed and the Seminal moisture, that is to say, Gold, our body by the reiteration of liquefaction in this our dissolving water, is turned and reduced into Sulphur and Argent-vive fixed, and so the perfect body of the Sun takes life in this water, is revived, inspired, increased and multiplied in his kind, as all other things are; for in this water it comes to pass that the body compounded of two bodies, of the Sun and of the Moon, puffs up swollen, putrifies as a grain of seed corn; becomes great with young, is lifted up and increases taking the substance and nature living and vegetable.

Also our water, or our foresaid Vinegar is the Vinegar of Mountain, that is to say, of the Sun and of the Moon and therefore it is mixed with the Sun and Moon and cleans them perpetually, to wit the body takes from this water the tincture of whiteness and with it (the water) shines with inestimable brightness: He therefore that knows how to turn the body into white Silver Mercurial, he may afterward by this white Gold, easily turn all imperfect metals into

very good and fine Silver; and this white Gold is by
the Philosophers called their white Moon. The white
Argent-vive fixed, the Gold of Alchemy and the white
smoke, therefore without that, our Antimonial Vinegar,
the white Gold of Alchemy cannot be made, and because
in our Vinegar there is a double substance of Argent-
vive, one of Antimony and another of Mercury sublimed.
It does therefore give a double weight and substance
of Argent-vive fixed, and also augment therein (in the
Gold) the natural colour, weight, substance, and
tincture thereof.

Therefore our dissolving water causes a great
tincture and a great fusion because that when it feels
the common fire, if there be in it the perfect body of
the Sun or of the Moon, it suddenly makes it to be
melted and to be turned into his substance, white as
it is and add colour, weight and tincture to the body,
it has power also to dissolve all things that may be
melted and it is a ponderous body viscous, precious
and honorable, resolving all crude bodies into theit
perfect matter, that is, into earth and a viscous
powder, that is to say, into Sulphur and Argent-vive.

If therefore you put into this water any metal
filed or attenuated and leaving it for a time in a
gentle heat, it will all be dissolved and change into
a viscous water or white oil, as is said, and so it
mollifies the body and prepares it for the fusion and
liquefaction, nay it makes all things fusible, that is

stones and metals and afterwards gives them life and Spirit, therefore it dissolves all things with a wonderful solution, turning the perfect bodies into a fusible medicine. Melting, penetrating and more fixed, increasing the weight and colour. Work therefore with it and you will obtain from it that which you desire, for it the Spirit and Soul of the Sun and Moon it is the Oil.

Dissolving Water, The Fountain, The Balneum Mariae, The fire against Nature, the moist Fire, The Secret Hidden and Invisible Fire and the most Sharp Vinegar, of which a certain Ancient Philosopher said; I besought the Lord and he showed me a certain clean water, which I know to be the pure Vinegar, altering and piercing and digesting; the Vinegar, I say, grenefiafiue and the instrument moving the Gold or the Silver to putrefie, resolve, and to be reduced into his First Matter; and it is the only agent in the whole world for this Art that can resolve and reincrudate or make raw again the metallick bodies with the conservation of their Species and without destruction unless it be t a new, more noble and better form or generation, that is to say, into the Perfect Stone of the Philosophers, which is their wonderful and hidden secret.

Now this water is a certain middle substance, clear as pure Silver, which ought to receive the Tinctures of the Sun and Moon, to the end that it may be congealed and converted into white and living earth, for this water has need of the perfect bodies, that with them after dissolution it may be congealed, fixed and coagulated into white earth; and their solution is their congelation, for they have one and the same operation, for the one is not dissolved but that the other is congealed, neither is there any other water which can dissolve the bodies but that which abides with them in matter and form, nay, it cannot be permanent except it be of the nature of the other body, that they may be made one together, therefore when you see the water coagulate itself with the bodies that be dissolved therein, rest assured that your Science, method and operations are true and Philosophical, and you proceed aright in the Art: Nature is amended in its like nature, that is Gold and Silver is amended in its like nature, that is Gold and Silver are amended in our water, also with the bodies, which water is called the mean of the Soul, without the which we can do nothing in this Art and it is the Vegetable, Animal and Mineral Fire preserving the Fixed Spirit, of the Sun and Moon the destroyer and conquoror of bodies because it destroys, dissolves and changes bodies and metallick forms and makes them to be no bodies, but a Fixed Spirit and turns them into a moist, soft and fluid substance which has ingression and power to enter into other imperfect bodies and so mixes with them by the smallest parts and so colours them and makes them perfect, which they could not do when they were metallick bodies, dry and hard, which have no entrance nor power to color and make perfect, imperfect bodies and therefore to good purpose do we turn the bodies into a fluid substance, because every tincture will colour a thousand times more when it is in a soft and liquid substance, then when it is in a dry one, as appears by Saffron and consequently the transmutation of imperfect bodies is impossible to be done by perfect bodies while they are dry, except they be first brought back into their First Matter, soft and fluid: From hence we conclude that we must make the moisture return and so return and so reveal that which is hidden, which is called the reincrudation or the making raw again of the bodies, that is, the boiling and the softening them until they be deprived of their hard and dry corporality or bodiliness, because that which is dry does not enter nor colour any more then itself; therefore the dry earthy body does not join except it be joined, because as I above said, that which is thick and earthy enters not nor colours, and because it enters not therefore it alters not, wherefore Gold colours not until the hidden Spirit be drawn from the belly.

Therefore our white water and that it be made altogether a Spiritual and white fume, the white Spirit and the wonderful Soul wherefore we ought by our water to attend, alter and soften the perfect bodies that they may afterwards be mixed with the other imperfect bodies and therefore if we had no other profit by that Antimonial water then this that it makes the body subtil, soft and fluid according to his own nature, yet it were sufficient for us, for it brings back the bodies to their First Original of Sulphur and Mercury, that of these we may afterwards in a short time, in less than one hour of the day do that above ground which Nature wrought underground in the mines of the earth in a thousand years, which is as it were miraculous and therefore our final secret is by our water to make the body volatile, spiritual and a joining water which has ingression or entrance into the other bodies, for it makes the bodys to be a very Spirit because it does increase (that is bring to the temper and consistence of wax) the hard and dry bodies and prepare them to fusion, that it turn them into a permanent or a bideding water, it makes then of the bodies a most blessed Oil, which is the true tincture and the white permanent water of Nature, hot and moist, temperate, subtil and fusible as wax, which pierces, reaching to the bottom, colouring and makes perfect; therefore our water does incontinently dissolve Gold and Silver and makes them an incombustible Oil which may then be mixed with other imperfect bodies, for our water turns the bodies into the nature of fusible Salt, which is by the Philosophers called Sal abroc, which is the best, noblest of all Salts being in the regiment thereof fixed and not flying the fire and it is indeed an Oil of a nature hot, subtil, penetrating, reaching to the depth and entering, called the compleat Elixir and it is the hidden secret of the wise Alchemist; he therefore that knows this Salt of the Sun and Moon and the preparation, thereof and afterwards how to mix it and make it friendly to the other imperfect bodies, he in truth knows one of the greatest secrets of Nature and one way of perfection. These bodies thus dissolved by our water are called Argent-vive which not without Sulphur, nor Sulphur without the nature of the Luminaries (or lights) because the lights (the Sun and Moon) are the means or middle things by which Nature passes in the perfecting and accomplishing the generating thereof and this quicksilver is called the Salt honoured and animated and pregnant (or great with child) and fire seeing that it is nothing but fire, nor fire but Sulphur, nor Sulphur but quicksilver drawn from the Sun and Moon by our water and reduced to a Stone of great price, that is to say, it is maker of the lights altered from baseness unto nobleness.

Note, that this white Sulphur is the Father of Metals and their Mother, together it is our MERCURY and the minera of Gold and the solve and ferment and mineral virtue and the living body and the perfect Medicine. Our Sulphur and our Quicksilver, that is, Sulphur of Sulphur and Quicksilver of Quicksilver and MERCURY of MERCURY the property therefore of our water is that melts Gold and Silver and augments in them their native colours, for it turns the bodies from corporality to spirituality and this water it is, which sends into the body a white fume which is the white Soul, subtil, hot and of such fierceness, this water is also called the bloody Stone and it is the virtue of the Spiritual Blood without which nothing is done and the subject of all liquable things and the liquefaction agrees very well and cleaves to the Sun and the Moon, neither is it ever separated from them, for it is kin to the Sun and the Moon, but more to the Sun than to the Moon.

Note this well, it is also the mean of conjoining the Tinctures of the Sun and Moon with imperfect metals, for it turns the bodies into a true tincture to join the other imperfect metals and it is the water which whitens, as it is white which quickens as it is a Soul and therefore (as the Philosophers say) soon enters into the body, for it is a living water which comes to moisted its earth, that it may bud and bring forth fruit in his kind and time, as all things springing from the earth are engendered by the dew or moisture; the earth therefore buds not without watering and moisture; it is the water of May dew that pierces them like rain water whitens them and makes one new body of two bodies.

This water of Life being rightly ordered with his body, whitens it and turns it into his white colours for the water is a white fume and therefore the body is whitened by it, whiten the body then and burn your books and between these two, that is, between the body and the water, there is friendship, desire and lust, as between the male and female, because of the nearness of their like natures, for our second Living Water is called Azote washing the Laton, that is, the body compounded by the Sun and Moon, by our First Water; the Second Water is called the Soul of our dissolved bodies, of which bodys we have already tied the souls together to the one, that they may serve the wise Philosophers.

O how perfect and magnificent is this water, for without it the work could never be brought to pass, it is also called, The Vessel of Nature, The Belly, The Womb, The Receptacle of the Tincture, The Earth, and this is the Fountain in which the King and Queen wash themselves and the Mother which must be put and sealed in the Belly of her Infant, that is, the Son which preceeded from her and which she brought forth, and therefore they love one another as a mother abd son, and are easily joined together because they come from one and the same Root, and are of the same substance and nature and because this, the water of the vegetable life, and therefore it gives life and makes the dead body to vegetate, increase and spring forth and rise from death to life by solution and sublimation, in so doing the body is turned into a Spirit and the Spirit into a body and then is made amity, peace, concord and union between the contraries, that is, between the body and the sirit, which reciprocally change their natures which they receive and communicate to one another by the least parts, so that the hot is mixed with the cold, the dry with the moist and the hard with the soft and thus is there a mixture made of contrary natures, that is, of cold with hot and of moist with dry and admirable connection and conjunction of enemies. Then our dissolution of bodies which is made in this First Water, which is no other thing than killing the moist with the dry, because the moist is coagulated with the dry, for the moisture is contained, terminated and coagulated into a body or earth only by dryness.

Let therefore the hard and dry bodies be put in our First Water, in a vessel well shut, where they may abide till they are dissolved and ascend on high, and then they may be called a new body. The White Gold of Alchemy, The White Stone, The White Sulphur, not burning and the Stone of Paradise, that is, the Stone converts imperfect metals into fine white Silver, having this, we have also the Body, Soul and Spirit all together, of the which Spirit and Soul it is said that they cannot be drawn from the perfect bodies but by conjunction of our dissolving water, because it is certain that the thing fixed cannot be lifted up but by the conjunction of the thing volatile, the Spirit by the mediation of water and the Soul is drawn from the bodies and the body is made no body because at the same instant the Spirit with the Soul of the body mounts on high into the upper part which is the perfection of the Stone and is called sublimation. This sublimation (says Florentinus Catalanus) is done by things sharp, volatile and spiritual, which are of a Sulphureous and Viscious nature, which dissolve the bodies and make them to be lifted up into the air in Spirit and in this sublimation a certain part and portion of our said First Water ascends with the body joining itself with them, ascending and subliming into a middle substance, which holds of the nature of the two, that is, of the bodies and of the water and therefore it is called the corporeal and spiritual compound. Corfusto Cambar Ethelia Zandariach, the good Duenech, but properly it is called the Water Permanent because it flys not into (away?) the fire always adhering to commixed bodies that is, to the Sun and Moon and communicating into them a living Tincture incombustible and most firm, more noble and precious than the former which these bodies had because from hence forward, this Tincture can run as Oil upon the bodies, perforating and piercing with a wonderful fixation because this Tincture is the Spirit and the Spirit is the Soul and the Soul is the body because in this operation the body is made a Spirit of a most subtil nature and likewise the Spirit is incorporated and is made of the nature of the body with bodies and so our Stone contains a body, a soul, and a spirit, which you could not do it if the Spirit was not incorporated with the bodies and the bodies with Spirits and made volatile or flying and afterwards permanent or abiding, therefore they have passed into one another and are turned the one into the other by wisdom. O Wisdom! How you make Gold to be volatile and fugitive, although by nature it must be fixed, it behoves then for to dissolve and melt those bodies by our water and to make them a permanent water, a Gold water sublimed, leaving in the bottom the gross earthy superfluous dry, and in this sublimation the fire ought to be soft and gentle, for in this sublimation the bodies be not purified in a lent or slow fire and the gross or earthly parts (note well) separated from the uncleanness of the dead, you shall be hindered from ever making your work perfect, for you need only this subtil and light nature of the dissolved bodies which our water will easily give you if you proceed with a slow fire, for it will separate the heterogeneous all (or that which is another kind) from the homogeneal (or that which is all one kind).

Our compound therefore receives mundification or cleansing by our moist fire, that is to say, dissolving and subliming that which is pure and white and casting aside the foeces like a voluntary vomit (says Asinaben) for such a disposition and natural sublimation, there is made a loosening or an untying of the Elements, a cleansing and separation of the pure from the impure, so that the pure white ascends upwards and the impure and earthly fixed remains in the bottom of the water or the vessel, which must be taken away and removed, because it is of no value, taking only the middle white substance, flowing and melting and leaving the feculent earth which remained below in the bottom, which came principally from the water and is the dross of the damned earth, which is nothing worth nor can ever do any good, as done the pure clear white and clean and after which we ought only to take, and against this Coepharean Rock the ship and knowledge of the Scholars and Students in Philosophy (as it happened also to me sometimes) most improvidently dashed and broken because the Philosophers do very often affirm the contrary, namely, that nothing must be removed or taken away, but the moisture, that is, the blackness which notwithstanding, they say and write only to deceive the unwise, gross and ignorant, which of themselves without a Master, unwearied reading or prayer unto God almighty, would like conquerors carry away the Golden Fleece. Note therefore that this separation, division and sublimation is without doubt the Key of the whole work, after the putrefaction then dissolution of these bodies. Our bodies do lift themselves up to the surface of the dissolving water in the colour of whiteness, the Antimonial and Mercurial Soul is by the appointment of Nature infused with the Spirits of the Sun and Moon, which separates the subtil from the thick and the pure from the impure, lifting up little and little: The subtil part of the body from the dregs, until all the pure be separated and lifted up, and in this is our Philosophical and natural sublimation fulfilled and this whiteness is the Soul infused into the body, that is, the Mineral virtue which is more subtil than fire, being indeed the true Quintessence and Life which desires to be born and to put off the gross earthy foeces which it has taken from the menstruous and corrupt place of his original and in this our Philosophical sublimation, not in the naughty common mercury which has no qualities like unto them wherewith our Mercury drawn from his Vitriolate Cavern is adorned, but let us return to our sublimation.

It is therefore most certain in this Art that this Soul drawn from the body cannot be lifted up but by the putting to a volatile thing which if of his own by which the bodies are made volatile and spiritual, lifting up, subtilating and subliming themselves against their own proper nature which is bodily and ponderous and by this means they are made no bodies but incorporeal and a Fifth Essence of the nature of the Spirit which is called Hermes, his bird and Mercury drawn from the red servant and so the earthly parts remain below or rather the grosser parts of the bodies which cannot by any wit or devices of man be perfectly dissolved and this white fume, this white gold, that this Quintessence is called also the compound Magnesia, which as a man contains or like a man is compounded of a body, a soul and a spirit, for the body is the fixed earth of the Sun which is more than most fine ponderously lifted up by the force of our divine waters, the Soul is the Tincture of the Sun and of the Moon proceeding from the conjunction or communication of these, but the Spirit is the mineral virtue of the two bodies and of the water which carries the Soul or the white tincture upon the bodies and out of the bodies as the tincture of dyers is carried by water upon the cloth and that Mercurial Spirit is the bond tyall of the Soul of the Sun and the body of the Sun is the body of fixation containing with the Moon the Spirit and Soul of the Spirit, therefore pierces the body, fixes the Soul, couples, colours and whitens; of these three united together in our Stone made, that is, of the Sun and Moon and Mercury, then with our gilded (or golden) water is extracted a nature surpassing all nature and therefore except bodies be by this our water dissolved, imbibed, ground, softened and sparingly and diligently governed until they leave their greatness and thickness and be turned into a thin and impalpable Spirit: Our labour will always be in vain, for unless the bodies be changed into no bodies, that is, into the Philosophers Mercury, the rule of Art is not yet found and the reason is because it is impossible to draw out of the bodies the most thin or subtil Soul, which has in it all tinctures, if the bodies be not first dissolved in our water. Dissolve therefore the bodies in the golden water and boil them until by the water all the tincture comes out into a white colour or white oil and when you shall see this whiteness upon the water, then know that the bodies are dissolved or melted and continue the decoction until they bring forth the cloud which they have conceived dark black and white, put therefore the perfect bodies in our water, in a vessel Hermetically sealed, upon a soft fire and boil them continually until they are perfectly resolved into a most precious oil.

Boil them (says Adfar) with a gentle fire as it were, for the hatching of chickens, until the bodies be dissolved and their tincture most nearly conjoined (mark well) be wholly drawn out, for it is not drawn out all at once, but it comes forth by little and little every day and every hour until after a long time this dissolution be compleat and which is dissolved do always arise uppermost upon the water, and in this dissolution let the fire be soft and continual until the bodies be loosened into a Viscous impalpable water and that the whole tincture come forth first in the colour of blackness, which is a sign of true solution, then continue the decoction until it become a white permanent water, for governing it in its bath it will afterwards be clear and is become like common Argent-vive climbing through the air upon first-and therefore when you see the bodies dissolved into a viscous water then know they are turned into a vapour and that you have the Souls separated from the dead bodies and by sublimation brought into the order and estate of Spirits, whereupon both of them with a part of our water are made spirits flying and climbing into the air and that there the body compounded of the male and female, of the Sun and Moon and of that most subtil nature cleansed by sublimation, takes life inspired by his moisture, that is by his water as a man by the air and therefore from henceforth it will multiply and increase in his kind, like all other things and therefore in such an operation and Philosophical sublimation, they are joined one with another and the new body inspired by the air, lives vegetably, which is a wonder wherefore unless the bodies be subtilized and made thin by fire and water until they do arise like spirits and be made like water and fume or like Mercury, there is nothing done in this art, but when they ascend they are borne in the air and are cleansed (changed) in the air and are made life with life in such sort that they can never be separated, as water mixed with water and therefore it is wisely said that the Stone is borne in the air because it is altogether spiritual, for the Vulture flying without wings crys upon the top of the mountain, saying, I am the white of the black and the red of the white and the citron of the red. I tell truth and lie not.

It suffices you therefore to put the bodies in the vessel and in the water, once for all and to shut the vessel diligently, until a true separation is made, which by the envious is called conjunction, sublimation, assation, extraction, putrifaction, ligation, dispensation, subtilation, generation, etc., and that the whole mastery be done, do therefore as in the generation, etc., and that the whole mastery be done, do therefore as in the generation of a man and every vegetable, put the seed once into the womb and shut it well, by this means you see that you need not many things and that our work requires no great charges because there is but one Stone in Medicine, one vessel, one regiment and one successive disposition to the white and to the red and all that we say in many places, take this, take that, yet we understand that it behoves to take but one thing and put it once into the vessel and to shut the vessel until work be perfected, for these things are set down by the envious Philosophers to deceive the unwary, as is aforesaid, for it is not this Art Cabalistical and full of Secrets; and do you fool, believe that we do openly teach the Secret of Secrets, and do you take our words according to the literal sound?

Know assuredly (I am no whit envious as others are) he that takes the words of the other Philosophers according to the ordinary signification and sound of them he does already having lost Ariadines thread, wonder in the midst of the Labyrinth and has as good as appointed his money to perdition.

But I Artephius, after I had learned all the Art and perfect science in the books of the true speaking Hermes, was sometimes as envious as all the rest, but when I had by the space of 1000 years or thereabouts, which are now passed over me, since my Nativity by the only grace of God almighty and the use of this wonderful Fifth Essence, when I say, for so long time I had seen no man that could work the Mystery of Hermes, by reason of the obscurity of the Philosophers words, moved with pity and with the goodness becoming an honest man, I have determined in these last times of my life to write all things, truly and sincerely, that you may want or desire nothing toward or to the Perfection of the Philosophers Stone (excepting a certain thing which it is not lawful for any person to say or write because it is revealed by God or by a Master and yet in this book he that is not shipwrecked, shall with a little experience easily learn it.

I have therefore in this book written the naked truth, although clothed with a few colours, that every good and wise man may from this Philosophical Tree happily gather the admirable apples of the Hesperides; therefore praised be the most high God, which has put this benignity into our Souls and with a wonderful long old age has given as a true decoction of heart, wherewithal it seems unto me that I do truly love, cherish and embrace all men, but let us return unto the Art; surely our work is quickly dispatched, for that which the heat of the Sun does in a hundred years in the mines of the earth, for the generation of a metal (as I have often seen) our secret fire, that is, our fiery Sulphurous water, which is called Balneum Marie, works in short time and this work is no great labour to him that knows and understands it; neither is the matter so dear, considering a small quantity suffices, that it ought to cause any man to pluck back his hand because it is so short and easy, that it may well be called the work of women and the play of children.

Work then cheerfully (my son) pray to God, read books continually, for one book opens another; think of it profoundly, fly all things that vanish in the fire, for you have not your infant in those combustible and consuming things, but only in the decoction of the water drawn from your ights, for by this water is colour and weight given infinitely and this water is a white fume, which as a Soul flows in the perfect bodies, taking wholly from them their blackness and uncleanness and consolidating the two bodies into one and multiplying their water, and there is no other thing that can take away their true colour from the perfect bodies, that is, from the Sun and Moon but Azoth, that is this our water which colours and makes white the red body, according to the regements thereof.

But let us speak of fires, our fire therefore is mineral, equal, continual, it vapours not unless it be too much stirred up, it partakes of Sulphur, it is taken otherwise then from the matter, it pulls down all things, it dissolves, congeals, and calcines, it is artificial to find, it is a short way (or an expense) without cost, at least without any great cost, it is moist, vaporous, digestive, altering, piercing, subtil, airy, not violent, not burning, compassing or environing, containing but one and it is the fountain of living water which goes and contains the place where the King and Queen bathe themselves. In all the work this moist fire is sufficient for thee at the beginning, midst and end, for in it consists the whole Art.

This is the fire natural against nature, unnatural and without burning, and finally this fire is hot, dry, moist and cold, think you upon this and work aright taking nothing that is of a strange nature, and if you do not well understand these fires, hearken further to what I shall give you, never as yet written in any book, for out of the abstruse and hidden cavitation of the ancients concerning fires.

We have properly three fires without the which the Art cannot be done, and he that works without them takes a great deal of care in vain. The first is the fire of the Lamp, which is continual, moist, vaporous, airy and artificial to find. For the Lamp ought to be proportioned to the closure (or inclosure) and herein we must use great judgement which comes not to the knowledge of a workman stiffneck, for if the fire of the Lamp be not geometrically and duly proportioned and fitted to the furnace, either for lack of heat you will not see the expected signs in their times and so you will lose your hope, by too long expectation or else with too much heat you will burn the flowers of the Gold and so sadly bewail your lost labour.

The second is the fire of ashes, in which the vessel hermetically sealed is shut up, or rather it is that most gentle heat which proceeding from the temperate vapour of the Lamp goes equally round about your vessel. This fire is not violent, if it be not too much stirred up, it is digesting, altering, it is taken from another body than the matter, it is but one, or alone, it is moist and unnatural etc.

The third is the natural fire of our water, which for this cause is also called fire against nature, because it is water and yet nevertheless it makes a meer spirit of Gold, which common fire cannot do. This fire is mineral, equil and partakes of Sulphur, it breaks, congeals, dissolves and calcines all. This is piercing, subtile, not burning and it is the fountain of living water wherein the King and Queen bathe themselves, whereof we have need in the whole work in the beginning, middle and ending; but the other two abovesaid we do not always need but only sometimes.

Join therefore in the reading of the book of Philosophers, these sorts of fires and without doubt you will understand their cavillations concerning their fires.

As touching the colours, he that does not made black cannot make white, because blackness is the beginning of whiteness and a sign of putrifaction and alteration and that the body is now pierced and mortified, therefore in the putrifaction, in this water, there first appears blackness like unto the broth wherein blood or some bloody thing is boiled. Secondly, the black earth by continual decoction is whitened because the Soul of the true bodies swims aloft upon the water, like white cream, and in this only whiteness all the spirits are so united that they can never fly from one another and therefore the laton must be whitened and fear the books lest our hearts be broken, for this entire whiteness is the true Stone, to the white and the body enobled by the necessity of his end and the tincture and whiteness of a most exuberant reflection and shining brightness, which being mixed with a body never seperates from it.

Here then note that the spirits are not fixed but in the white colour, which by consequence is more noble than the other colour and ought more earnestly to be desired. Considering it is as it were the complement and perfection of the whole work, for our earth is first putrified into blackness, then is cleansed in the elevation or lifting up, afterwards being dried, the blackness then it is whitened and the dark, moist domination of the woman perishes and then the white fume pierces into the body and the spirits are shut up or bound together in dryness, or that which is corrupting, deformed and black with moisture vanishes and then the new body arises again, clear white and immortal, getting the victory over all his enemies and as heat working upon that which is moist, causes or engenders blackness, which is the first colours, so by decoction ever more and more heat working upon that which is dry, begets whiteness which is the two colours and afterwards working upon that which is purely and perfectly dry, it causes citrinity of redness, of so much concerning the colours.

We must therefore understand that the thing which has the head red and white, his feet white and afterwards red and yet before that, the eyes black, this only is our mystery, dissolve then the Sun and the Moon in our dissolving water, which is familiar, friendly and of the next nature unto them, which is likewise to them sweet and pleasant and as it were a womb, a mother, an original, the beginning and the end of life and that is the reason why they are amended in this water, because that nature rejoices in nature and nature contains nature and in true marriage they are joined together and made one nature, one new body raised up and immortal and this we must join consanguinity of those natures, will meet and follow one another, putrify themselves, engender themselves and make one another rejoice, because nature is governed by nature, which is nearest and most friendly to it. Our water (says Danthin) is the most pleasant, fair and clear fountain, prepared only for the King and Queen, whom it very well and they know it, for it draws them to itself and they abide therein, to wash themselves 2 or 3 days, that is, 2 or 3 months, and it makes them young again and fair and because the Sun and Moon have their original from this water, their mother, therefore it behooves that they enter again into their mother's womb, that they may be born again and made more strong, more noble and more valiant and therefore if these do not die and not be turned into water, they remain alone and without fruit, but if they die and be dissolved in our water, they bring fruit a hundred fold and from that very place where it seemed that they had lost what they were, from thence shall they appear that which they were not before, let therefore the spirit of our living water be with great and subtility fixed with the Sun and Moon because they turned into the nature of water, do die and seem ike unto the dead. Yet afterwards being inspired from thence, the live increase and multiply like all other vegetable things.

It is enough then to dispose the matter sufficiently from without, for from within itself does work sufficiently to its own perfection, for it has in itself a certain and inherent motion, according to the true way better than any order that can be imagined by man and therefore do you only prepare and nature will perfect, for if she be not hindered by the contrary, she will not pass her own certain motion, on well to conceive as to bring forth, wherefore after preparation of the matter, take heed lest by too much fire you make bath too hot, take heed lest the spirit do exhale, because it would hurt him that works, that is to say, it would destroy the work and cause many infirmities (that is) much sadness and anger, from this that has been spoken is drawn the axiom, to wit, that by the course of nature he does not know the making of metals, that knows not the destruction of them, it behooves then to join together them that are of kindred, for natures do find their like natures and being putrified are mixed together and mortify themselves; it is necessary therefore to know this conjunction and generation and how the natures do embrace one another and are pacified in a slow fire, how nature rejoices in nature and nature retains nature and turns into a white nature, after this, if you will make it red you must boil this water in a dry

126

continual fire, until it be as red as blood which will be nothing else but fire and a true tincture and so by a continual dry fire the whiteness is changes, amended, perfected and made citrine and acquires redness, a true fixed colour and consequently by howmuch the more this red is boiled, so much the more it is coloured and made a tincture of perfect redness, wherefore you must with a dry fire and a dry calcination, without any moisture, boil this compound until it be clothed with a most red colour and then it will be a perfect elixir.

If afterwards you will multiple it, you must again resolve that red new dissolving water and after by decoction, whiten and rubified it by the degrees of fire, reiterating the first regiment, dissolve, congeal, reiterate, shutting, opening and multiplying in quantity and quality at thy own pleasure.

For by a new corruption and generation there is again brought in a new motion and so we could never find an end, if we would always by reiteration of solution and coagulation by the means of our dissolving water, that is to say, dissolving and congealing, as is said in the first regiment and so the virtue thereof is increased and multiplied in quantity and quality, so that if in the first work one part of your Stone will join a hundred, in the second it will join a thousand, in the third ten thousand and so by pursuing your work your projection will come to

127

infinity, joining truly and perfectly and fixedly, every quantity how great soever it be and so by a thing of an easy price is added colour and virtue and weight, therefore but our fire and azoth are sufficient for you, boil, boil, reiterate, dissolve, congeal and so continue according to your will, multiplying it as much as you will and until your Medicine is made fusible as wax and that it have the quantity and virtue which you desire, therefore all the accomplishment of the work or of our 2 Stones (note it well) consists in this, that you take the perfect body which you must put in our water, in a house of glass well shut and stopped with cement, lest the air get in, or the moisture enclosed get out and there hold it in the digestion of a gentle heat, as it were of a bath, or the moist temperature of dung, upon which with the fire you shall continue the perfection of decoction until it be putrified and resolved (or dissolved) into black and afterwards be lifted up and sublimed by the water that it may thereby be cleansed from all blackness and darkness and that it may be whitened and made subtil, until it come to the utmost purity of sublimation and at last be made volatile and white within and without, for the Vulture flying in the air without wings cries that it may get upon the mountain, which is upon the water, upon which the white spirit is caused, then continue a convenient fire and your spirit, that is, the subtil substance of the body of Mercury will ascend upon the water, which quintessence is whiter than the snow, continue still and in the end strengthen your fire until all which is spiritual mount on high, for know well that all is clear, pure and spiritual ascends on high in the air in the form of a white fume, which the Philosophers call the Virgins Milk, it behoves therefore that (as Sybil said) the Son of the Virgin be exalted from thye earth and that the white quintessence after his resurrection, be lifted up towards heaven and that the gross and thick remain in the bottom of the vessel and of the water, for afterwards when the vessel is cold, you will find in the bottom thereof the foeces, black, burnt up, combust, separate from the spirit and white Quintessence, which dregs you must cast away.

In these times the Argent-vive rains from our Air, upon our new Earth which is called Argent-vive, sublimed from the Air whereof is made a water, viscous, clean and white, which is the true Tincture, separated from all black foeces and so our brass or laton is without water governed, purified and adorned with a white colour, which white colour is not gotten but by decoction and congelation of the water, boil it then continually, wash away the blackness from the laton, not with your hand but with the Stone or the fire or our second Mercurial water, which is the true Tincture for this separation of the pure from the impure, is not done with the hands, but nature herself alone by working it circularly to perfection brings it to pass, it appears then that this composition is not a manual work, but a change of nature, but becaure nature dissolves and conjoins itself, is sublimed and lifts up itself and having separated the foeces it grows white and in such a sublimation tha parts are always joined together more subtil, more pure and essential because that when the fiery nature lifts up the subtil parts, it lifts up always the more pure and by consequence leaves the gross in the bottom and therefore it behoves by an indifferent fire to sublime in a continual vapour, that the Stone may be inspired in the air and live, for the nature of all things takes life of the inspiration of air and also all our mastery consists in vapour and is the sublimation of water and therefore our brass or laton must by degrees of fire be lifted up and freely without violence of himself ascend on high, wherefore unless the body be by fire and water be dissolved, attenuated and subtilized until it ascends as a spirit or climb like Argent-vive or as the white Soul separated from the body and carried in the sublimation of the spirits, there is nothing at all done in this Art, but when it ascends on high is born in the air and changed in the air and is made, life being altogether spiritual and incorruptible, and so a Regiment the body is in such made a spirit of subtil nature and the spirit is incorporated with the body and is made one with it and in such sublimation, conjuction and elevation, all things are made white and therefore this Philosophical and Natural sublimation is necessary, for that it makes peace between the body and the spirit, which is impossible otherwise to be done then by this separation of the parts.

Wherefore it behooves to sublime them both, to this end, that in the troubles of this stormy sea the pure may ascend and the impure and earthly may descend and for this cause it must be boiled continually, that it may be brought to a subtil nature and that the body may assume and draw to itself the white Mercurial Soul, which it naturally retains and suffers it because it is like unto it in the nearness of the first pure and simple nature.

From hence it appears that this separation must be made by decoction until there remains no more of the fat of the Soul, which is not lifted up and exalted in the upper part, for so they shall be both reduced into a simple equality and unto a simple whiteness, the Vulture therefore flying in the air and the Toad going upon the earth, is our mastery and therefore when you shall gently and with great discretion separate the earth from the water, that is, the fire and the subtil. From the thick then, from that which is pure will ascend from earth into heaven and that which is impure will go down to the earth and the more subtil part will in the upper place take a nature of a spirit and in the lower place take the nature of an earthly body, wherefore let the white nature with the more subtil part of the body be by this operation lifted up, leaving the foeces which is done in a short time, for the Soul is aided by her associate and fellow and perfected by it.

My Mother (says the body) has begotten me and by me she herself is begotten and after she has taken her flight (or I have taken from her flying) she after the best manner, she can become a pious Mother, nourishing and cherishing the Son whom she has begotten until she come to a perfect state.

Hear this Secret; keep the body in this our Mercurial water until it ascends on high with the white Soul and the earthly descend to the bottom, which is called the earth that remains, then shall you see the water coagulate itself with its body and be assured the Science is true, because the body coagulates his moisture into dryness, as the rennet of a lamb coagulates milk into cheese; in the same fashion the spirit will pierce the body and there will be a perfect mixture made by the least parts and the body will draw into himself his moisture, that is to say, this white Soul, even as the lodestone draws the iron, because of the likeness and nearness of his nature and his greediness and then the one will hold the other and this is our sublimation and coagulation which retains everything volatile and makes that it can fly no more, therefore this composition is not a manual action, but (as I said) a changing of natures and a wonderful connection of their cold with hot and their moist with dry, for the hot is mixed with cold and their dry with moist and so by this Mercury is made the mixture and conjunction of the body with the spirit which is called the changing of contrary natures, because that in such a solution and sublimation the spirit is turned into a body and the body into a spirit, so that the natures being mingled together and reduced into one do change one another in as much as the body makes the spirit a body and the spirit turns the body into a joined and white spirit.

And therefore (this is the last time I will tell you) boil it in our white water, that is, in Mercury until it be dissolved into blackness and then by continual decoction it will be deprived of this blackness and the body so dissolved will at length arise with a white Soul and they will embrace one another, so that they will no more be divided asunder and the spirit is united to the body with a real accord and are made one permanent thing and this is the solution of the body and the coagulation of the spirit, which have one and the selfsame operation.

He therefore that knows how to marry, to make with child, to mortify, to putrify, to engender, to quicken the species, to bring it into the white light and to cleanse the Vulture from his blackness and darkness until he be purged by fire, coloured and putrified from all his spots shall be the owner of so great a dignity, that Kings shall reverence him and do him honour.

Wherefore let our body abide in the water until such time as it is loosened into a new powder in the bottom of the vessel and of the water which is called the black ashes and this is the corruption of the body which is by wise men called, Saturn, Laton or Brass, The Philosophers Lead and the discontinued powder and is the putrifaction and resolution of the body; there appears three signs, to wit, the black colour, the dis-continuity of the parts and a stinking smell, which is likened to the smell of a sephulcheur or graves; this ashes then is the of which the Philosophers have said so much, which remained in the lower part of the vessel, which we ought not to despise, for in it is the diadem of our King and Argent-vive, black and unclean, from whence the blackness must be purged by continual decoction in our water until it be lifted up in a white colour, which is called the Goose and the Poulet of Hermogenes.

He therefore that makes the red earth black and then white has the mastery, as also he that kills the living and quickens the dead, therefore make black, white and then the white red, that you may make your work perfect and when you see the true whiteness appear, which shines like a naked sword, know that in that whiteness is redness hidden and then you must not take out of the vessel that whiteness, but only boil it to the end, that which dryness and heat, there may come upon it a citron colour and in the end a most shining and most sparkling red, which when you see it, with great fear and trembling, praise the most good and most great God, which gives wisdom and by consequence riches to whom he pleases and according to the iniquity of persons, take them away again and deprives them of them forever, plunging them in the servitude and slavery of their enemies.

To Him be praise and glory forever and ever AMEN.

The End.

A Word from the Publisher

Thank you for purchasing this small work from The R.A.M.S. Library of Alchemy. During his lifetime, Hans Nintzel was dedicated to the identification, acquisition, study, retyping and, when necessary, translation of what he considered to be the most important known works on Alchemy. Hans was assisted by his sparse network of fellow Alchemists, all members of the Restorers of Alchemical Manuscripts Society (R.A.M.S.). I was an active member of R.A.M.S.

My goal is to publish all of the works originally made available through R.A.M.S. as photocopies. To facilitate this, I have chosen to have the books professionally printed. I also have a few titles that I intend to add to the original R.A.M.S. Library, selected by strict criteria established by Hans.

If you have a work on Alchemy that you believe should be a part of the R.A.M.S. Library, please contact me through R.A.M.S. Publishing Company.

Philip N. Wheeler

www.ingramcontent.com/pod-product-compliance
Lightning Source LLC
Chambersburg PA
CBHW080814180526
45168CB00006B/2442